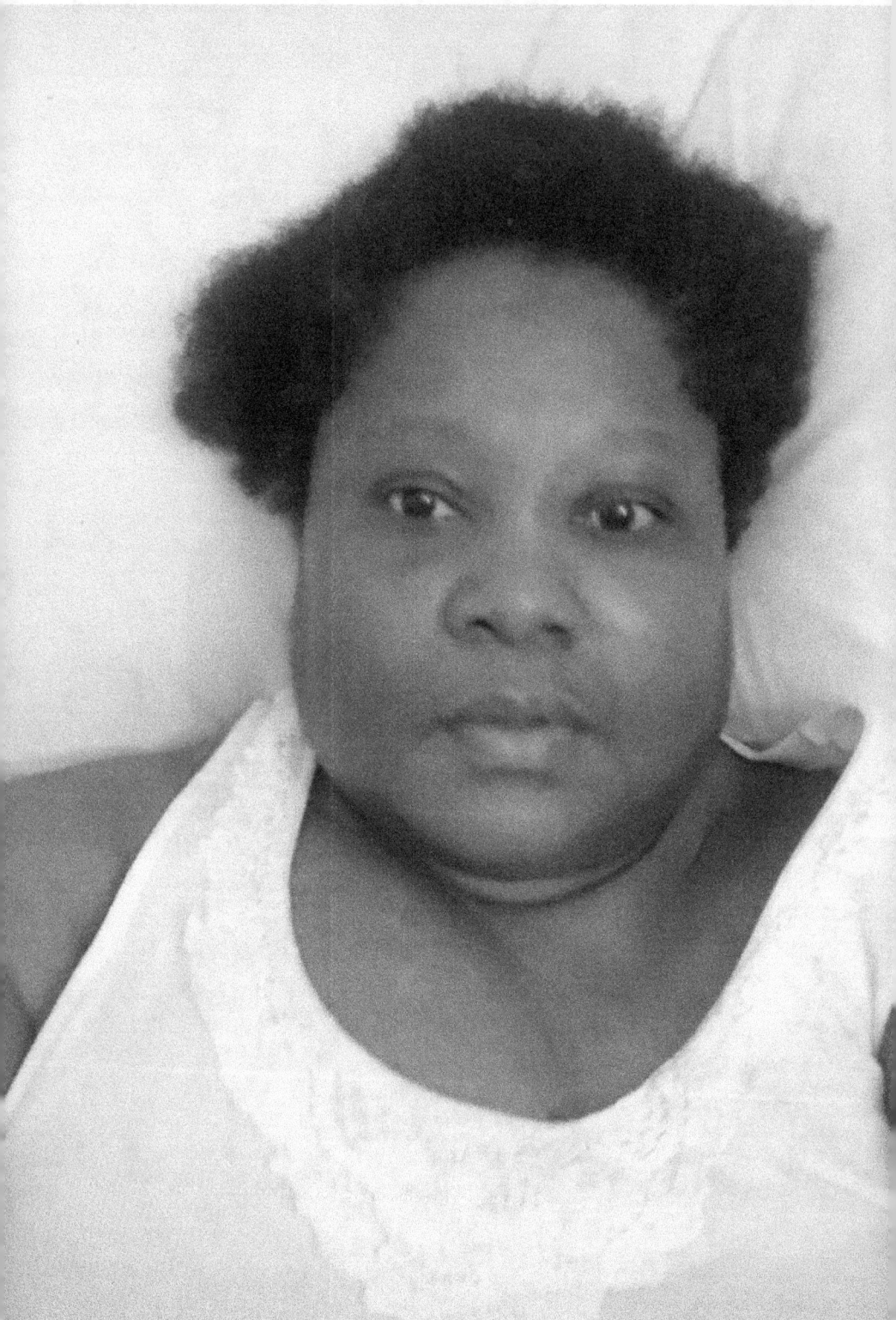

PM 2:16

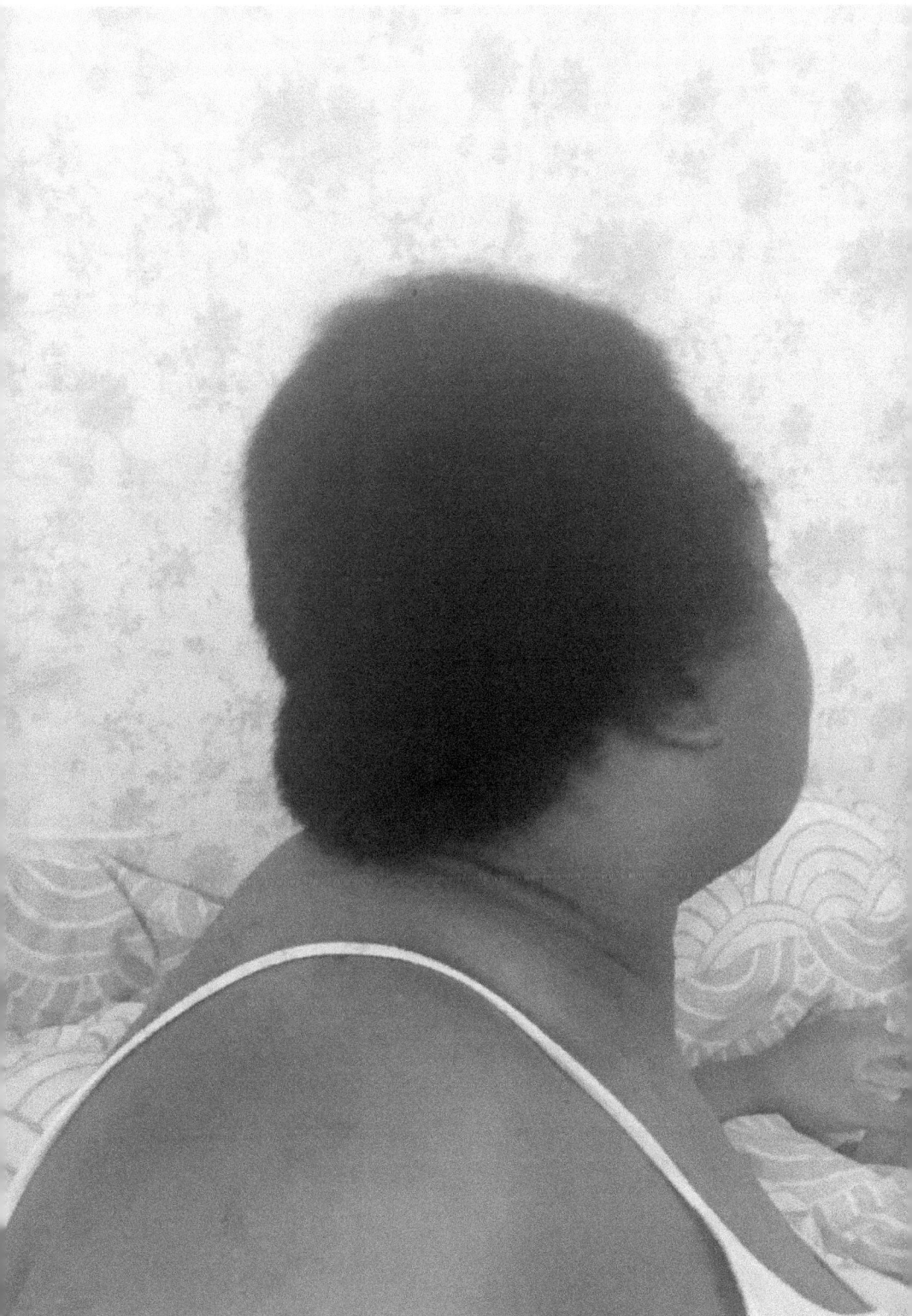

Start

natural
plastic surgery

Harry
Potter

March 21 2019 Continued

8:02pm: Brewer yeast x6 + tan x2
+ flaxseed x 3 + charcoal x 1
* + Calcium DK X 2

9:21pm: Calcium DK x1 + Brewer yeast x6

9:41pm: flaxseed x3 + bathroom

10:27pm: hydro collag x2

11:06pm: Omega 3 x3 + flaxseed x3
+ Brewer yeast x6

11:34pm: charcoal x1

Harry
Potter

March 22 2019

1:32am: Brewer yeast x6 +
charcoal x1 + St John's wort x1
+ flaxseed x4 + l-arginine x1 +

1:47am: Omega3 x4 + taur 1

3:51am: Brewer yeast x6 +
charcoal x1 + St John x1 + lysine x1
+ arginine x2 + taur x1 + omega3 x3
+ swollen under eye right
+ flaxseed x3 + ashwag x2
+ uracit x2 + exfo

5:15am: Brewer yeast x6 + uva x1
+ flaxseed x3 + ashwag x1

7:27am: Uva x1 + Brewer yeast x6
+ taur x1 + Omega3 x3

10.22am: lysinex1 + Brewers yeast x6
+ charcoalx1 + flaxseed x3 +
Omega3 x3 + l-arginine x2
+ taux1 + lysinex1

11.34am: Omega3 x4 + charcoalx1
+ uvaultix1

12.02pm: arginine x2 + flaxseed x3

12.11pm: Ironx1

Home by 9.44pm

10pm: Amoxicillin capsule x1
(prescribed)

10.45pm: flaxseed x1

11.23pm: Brewers yeast x6

12.14am [March 23 2019]

Charcoal x1 + tau x1 + Calcium DK x1
+ Brewers yeast x6 + Omega3 x3
+ amino1500 x1 + l-arginine x1

2.52am: Brewers yeast x6
+ Charcoal x1 + flaxseed x3
+ tau x1 + ashwag x1
+ Omega3 x3 + Iron x1

Harry
Potter

The prescribed
Amoxicillin boosts
the speed of herbs to
remove excess uric acid
and go off the boat

Harry
Potter

Walking without boot

5:33am: Brewer yeast x6 + uva x1
+ flaxseed x3 + charcoal x1
   gum pain

6:14am: Omega3 x3 + Brewer
yeast x6 + tan x1

7:03pm: Brewer yeast x6 +
   uva x1 + charcoal same
Puffy cheek, lip pain

7:17pm: uva x3 + ashwag x1

8:02pm: omega3 x3 + uva x5
+ Brewer yeast x6

9:14pm: Brewer yeast x6 +
charcoal x1 + flaxseed x3
+ uva ursi x5
Treating foot + mouth as/for
cancer treatment

11:49am: bathroom + ura x5
+ Brewers yeast x 6 + ashwag x 1 +
omega 3 x 3 + amoxicillin x 1
walking without boot

12:19pm: Brewers yeast
+ ura uti x5

1:33pm: flaxseed x 3 + ura x5
+ charcoal x 1 + omega 3 x 3

2:50pm: ura uti x 6 +
Brewer yeast x 6

3:06pm: amoxicillin x 1

4:42pm: Omega 3 x 3 + ashwag x
tau x 1 + ura x 5 + lysine x 1
+ Brewer yeast x 6 + Calcium D K x 2

5.36pm: flaxseed×3
6:01pm: flaxseed×3 + Brewers yeast ×6

6.34pm: flaxseed×3
6.58pm - flaxseed×3 teeth discomfort
7.21pm: Charcoal×1 + Brewers yeast×6
+ flaxseed×3 + uva×4

7.50pm: flaxseed×3 + Brewers yeast ×4

Stop Coca-Cola for now

8.18pm: flaxseed×3 + uva×4

8.48pm: ashwag×1

9.19pm: Brewers yeast×6 +
charcoal×1 + menopause×1

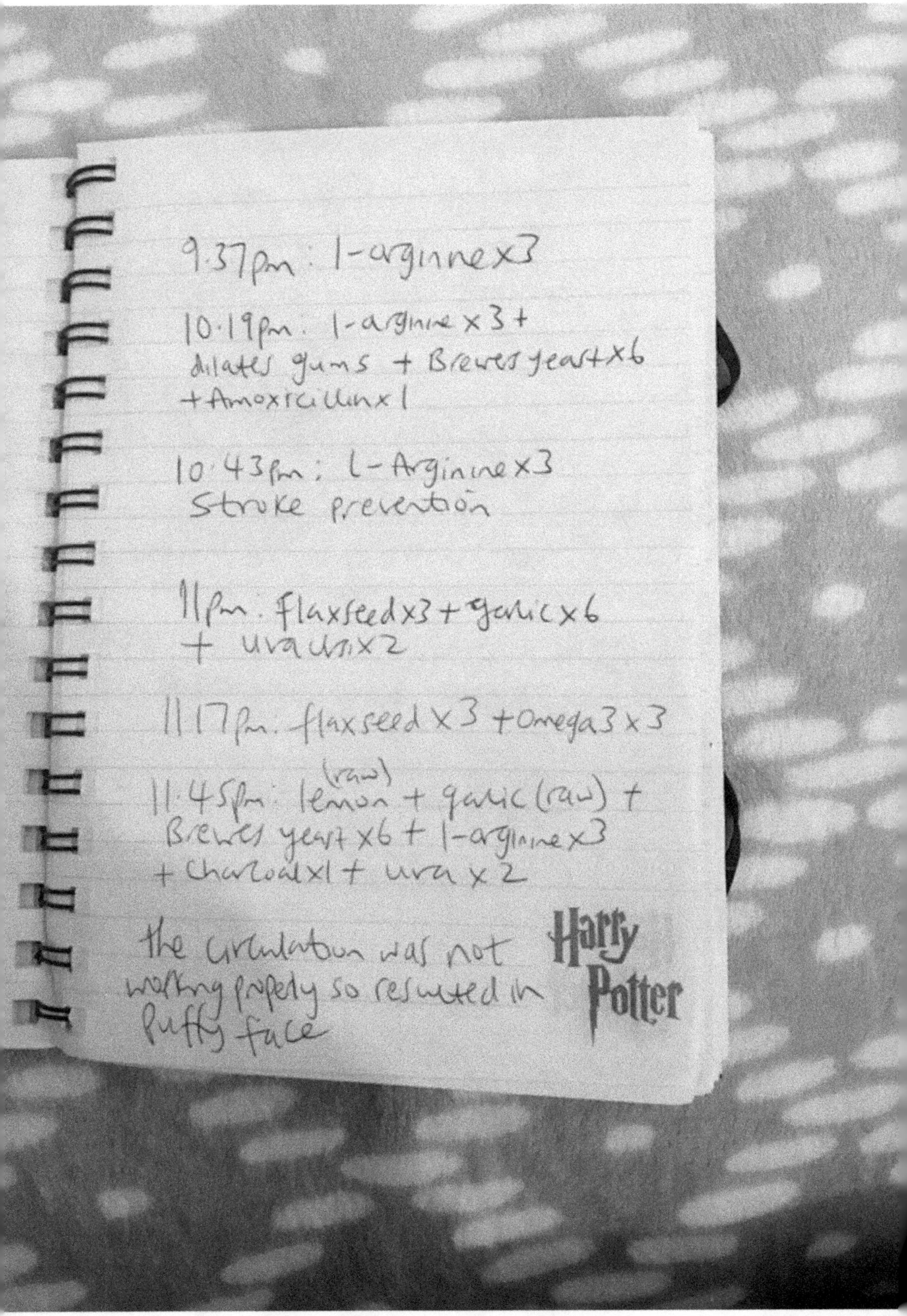

9.37pm : l-arginine x3

10.19pm : l-arginine x3 +
dilates gums + Brewers yeast x6
+ Amoxicillin x1

10.43pm : L-Arginine x3
Stroke prevention

11pm : flaxseed x3 + garlic x6
+ uva ursi x2

11.17pm : flaxseed x3 + omega 3 x3

11.45pm : lemon + garlic (raw) +
Brewers yeast x6 + l-arginine x3
+ charcoal x1 + uva x2

the circulation was not
working properly so resulted in
puffy face

Harry Potter

12:41 am [march 24 2019]

Brewer yeast x6 + L-arginne x3
+ wax2 + flaxseed x3

1:15am flaxseed x3

1:23 am: L-arginne x3 + Omega3 x3

2:47 am: flaxseed x3 + bathroom +
Brewer yeast x6 + wax3
+ charcoal x1 + arginne x3
+ mouth eye pain. + Omega3 x1
extremely acidic stiff teeth

3:20am: aminolsuu x1 +
flaxseed x3

4:04 am: aminolsuu x2 +
flaxseed x3 + Brewer yeast x6

6.16am: Brewers yeast x6
+ amino 1500 x1 + ura x 2 + flaxseed x3

8.43am: Brewers yeast x6 +
flaxseed x3 + ura x 2
+ amino 1500 x 1

10.45am: Brewers yeast x6 +
flaxseed x3 + l-arginine x3 +
amino 1500 x1 + Charcoal x 1

1.24pm: Brewers yeast x6 + ura x 2
flaxseed x3 + amino 1500 x1 + Omega 3 x 3

2.24pm: amoxicillin x1
+ Brewers yeast x6

Harry
Potter

flaxseed oil cleanses the teeth + neutralizes acid s

Harry Potter

3.57pm: L-arginine x3 + taux1
+ Brewers yeast x6 + flaxseed x3
+ omega3 x3 + ura x2 + aminolsax1

5.19pm: Brewers yeast x6 +
flaxseed x3 + ura x2 + aminolsax1

6.13pm: St John's wort x1
+ omega3 x3 + flaxseed x3

6.56pm: ashwag x1

7.26pm: omega3 x3

7.57pm: omega3 x3 +

8.24pm: omega3 x3
gum melting (inside mouth sore)
8.47pm: omega3 x4

9.22pm: omega3 x4
+ flaxseed x3 ankle (back) pain
left foot + urax1 x2

10.05pm: omega3 x4

11.07pm: Omega3 x4 + urax1

11.28pm: omega3 x4 +
Charcoal x1 + flaxseed x3

12.56am / March 25 2019

flaxseed x3

1.08am: Brewers yeast x6

7.26am: Brewers yeast x6 + uri x1
+ flaxseed x3 + Omega3 x3
+ argnine x2

8:22am: flaxseed x3
+ Brewers yeast x6

12:32pm: flaxseed x3 + uva x2
+ omega3 x3 + Brewers yeast x6
l-arginine x2 + Charcoal x1
+ ashwag x1 + aminolsuvs x1

12:51pm: flaxseed x3

1:38pm: flaxseed x3 + Brewers yeast x6
+ full body toptail exfol. shower

2:02pm: Brewers yeast x6
+ omega3 x4 + uva x2

2:33pm: Brewers yeast x6 +
omega3 x3 + flaxseed x3

3.50pm : omega3 x 3 + flaxseed x 3

4.49pm : flaxseed x 3 + Charcoal x 1
+ Brewers yeast x 6 + omega3 x 3
+ Calcium D K x 1

5.56pm : L-arginine x 3
some Cough

6.32pm : aminos 00 x 1

7.10pm : Omega 3 x 4 + flaxseed x 3
+ Brewers yeast x 6

8.27pm : Omega 3 x 3 + Calcium D K x 1

9.18pm : Charcoal x 1 + lion x 1
+ Brewers yeast x 6 + lysine x 1
+ omega 3 x 4

Harry Potter

10:08 pm: L-arginine x3 + ashwag x1 + Brewer yeast x6

11:32 pm: Brewer yeast x6 + flaxseed x3 + uva x2

2:14 am (March 26 2019)
Brewer yeast x6 + uva x1 + omega3 x3

Charcoal x1

4:01 am: flaxseed x3 + uva x1 + Brewer yeast x6 + L-arginine x3

4:29 am: bathroom + ashwag x1

9:40 am: Brewer yeast x6 + amino1500 x1 + Charcoal x1 + omega3 x3 + uva x2 + flaxseed x3

Harry Potter

bus 266 → Petrol Station
bus stop → need the bathroom
→ bus 266 → Sainsburys
for napkins, veg cube, soup, cake

→ bus 440 → Home →
bathroom

Harry
Potter

home by 11.40am

home by 4.19pm

bus 440 → police bus stop →
bus 207/427 → Ealing Bdw →
haB uva, ever firm, organic white tea/
fennel/peppermint tea & coconut water
+ l-arginine + 5HTP + sage
    → ashwag pukka → primark towels
top, blanket, bedsheet, pillowcase, duvet
cover → marksons supermart etc.
→ bus 207/427 → tesco action bus stop → bus 440 → home

4.46pm: Iron tonic + uva x3 +
    brewers yeast x6 + 5htp x1 +
even prim oil x3

5.34pm: aminos 1500 x1 +
ashwag x1 + brewers yeast x6

Harry
Potter

6.01pm: stressed out
left foot + evening primrose
oil x3

7.25pm: Brewer yeast x6 + Urax2

8.39pm: Brewer yeast x6 + Urax2
+ Evening Primrose oil x3

9.22pm: Omega3 x3 + amino15av x1
+ Even primrose oil x3 + Brewer yeast x6
+ charcoal x1 + ashwag x1

1.21am [March 27 2019]
Brewer yeast x6 + Even Primoil x1
+ Omega3 x2 + Urax2

Harry Potter

CBD Cannabidiol tea
Lighters smaller foot like
iceblocks would but a
little more lighter.
CBD relieves stress

Harry
Potter

2.14
even

5.09
omega

6.

9.29
+ even

Home

1.53

2.21pm

3.10pm
tannou
+ even P

2.14am: Brewers yeast x 6 +
evening primrose oil x 2

5.09am: even. prim oil x 2 +
Omega 3 x 3 + Brewers yeast x 6

6.24am: flaxseed x 3

9.29am: Brewers yeast x 6
+ even prim oil x 2 + 5HTP x 1

Home by 1.35pm

1.53pm: tasting CBD tea

2.21pm: 5htp x 1

3.10pm: Charcoal x 1 + uva x 2
+ tannoisoox x 1 + Brewers yeast x 6
+ even prim x 3 + Calcium D K x 1

2.14 am: Brewers yeast x6 + evening primrose oil x2

5.09am: even prim oil x2 + Omega3 x3 + Brewers yeast x6

6.24am: flaxseed x3

9.29am: Brewers yeast x6 + even prim oil x2 + 5HTP x1

Home by 1.35pm

1.53pm: tasting CBD tea

2.21pm: 5htp x1

3.10pm: Charcoal x1 + uva x2 + amino acox1 + Brewers yeast x6 + even prim x3 + Calcium DK x1

Harry Potter

7.48pm: Brewers Yeast x6

8.10pm: cloudy eyes + evening
primrose oil x3 + omega3 x3 +
Brewers yeast x6

9.37pm: ashwag x1 + silica x1

10.16pm: Brewers yeast x6 +
ever prim x2

10.46pm: Brewers yeast x6
+ sage x4

1.58am (march 28 2019)
Brewers yeast x6 + uva x2 +
amino 1500 x1 + omega 3 x3
+ flaxseed x1

Harry Potter

2:38am: sage x5 + ashwag x1
+ omega3 x2

4:29am: even prim x1 + omega3 x2
+ ashwag x1 + Brewer yeast x6

4:56am: Sht? x1

7:33am: Brewer yeast x6 +
even prim x 2

9:01am: Brewer yeast x6 +
l-arginine x3 + even prim x 2

9:42am: Brewer yeast x6 +
Omega3 x2

Home delivery aries
                    arrives

Harry
Potter

12.07 pm: Brewers yeast x6
+ evea prim x2

12.46 pm: Silica x1 + Hydro Collag x1

1.21 pm: ashwag x1

2.02 pm: uva x1 + Brewes yeast x6

home by 6.36 pm

7.03 pm: ashwag x2 sneezed

7.27 pm: Charcoal x1 + menoface x1
+ 5HTP x2

8.20 pm: Organic white tea, fennel
+ Peppermint tea + Honey
Stiff neck and clavicle

Harry Potter

as the therapist put a towel
on my front foot plantar fascia
and towel on her right thigh,
she pushed the dorsal foot as if
to pump foot, thus, resulting
in clavicle pain afterwards as
side effect.

Harry
Potter

8.4
1-0

9.5
t

10.
W

12.

un

12.
but
pain

1.05c

8.47pm: stiff clavicle +
l-arginine x3 + even. prim oil x3

9.54pm: Brewers yeast x6
+ flaxseed x2

10.23pm: stroke tenderness +
uva x2

12.34am: march 29 2019
uva x2 + Brewers yeast x6

12.58am: l-arginine x3 +
bathroom + omega3 x3 + clavicle
pain

1.05am: ashwag x1

Harry
Potter

1·58am: flaxseed x1 + neck pain
+ Brewers yeast x6 + ura x2 +
amino 1500 x1 + Charcoal x1

3·43am: l-arginine x3 +
Brewers yeast x6 + omega3 x3

6·03am: ashwag x1 + lysine x1
+ ura x2 + flaxseed x1 + omega3 x4
+ l-arginine x3 + Brewers yeast x6
+ even prim oil x2

7·52am: flaxseed x1 + omega3 x4
+ l-arginine x3 + Brewers yeast x6
+ ashwag x1 + shtv x1 + ura x2
+ even prim x1 + menopace x1

1:58am: flaxseed x1 + neck pain
+ Brewers yeast x6 + ura x2 +
amino 1500 x1 + charcoal x1

3:43am: l-arginine x3 +
Brewers yeast x6 + omega3 x3

6:03am: ashwag x1 + lysine x1
+ ura x2 + flaxseed x1 + omega3 x4
+ l-arginine x3 + Brewers yeast x6
+ even prim oil x2

7:52am: flaxseed x1 + omega3 x4
+ l-arginine x3 + Brewers yeast x6
+ ashwag x1 + shtr x1 + ura x2
+ even prim x1 + menopale x1

Harry Potter

Pneumonic lip and nose
swelling = Omega 3 + Brewers
yeast

Walking barefeet is still
slow, improves with shoe scholls.

Harry
Potter

4. 4.
pain +
+char
Brew

5.5

6.4

7.43
+ fl

8.43
+ a
Pneumon
no 60
Pneumoni
and nos

4.45pm: amoxicillin x1 + clavicle pain + omega 3 x 4 + amino 1500 x1 + charcoal x1 + menopace x1 + Brewers yeast x6

5.58pm: uva x1

6.44pm: Brewers yeast x6

7.43pm clavicle stroke symptoms + flaxseed x4 + tau x1

8.43pm: L arginine x2 + silica x1 + Ashwag x1 + Brewers yeast x6
Pneumonic + vapor rub around neck
no boot worn today
Pneumonic little swell under lip and nose

Harry Potter

4.45pm: amoxicillin x1 + Clavicle
pain + Omega 3 x4 + amino/500 x1
+charcoal x1 + menopace x1 +
Brewers yeast x6

5.58pm: uva x1

6.44pm: Brewers yeast x6

7.43pm Clavicle stroke symptoms
+ flaxseed x4 + tau x1

8.43pm: L.arginine x2 + Silica x1
+ Ashwag x1 + Brewers yeast x6
Pneumonic + vapor rub wand neck
No boot worn today
Pneumonic little swell under lip
and nose

Harry Potter

4.45pm: amoxicillin x1 + clavicle
pain + omega 3 x4 + amino 1500 x1
+charcoal x1 + menopace x1 +
Brewers yeast x6

5.58pm: uva x1

6.44pm: Brewers yeast x6

7.43pm Clavicle stroke symptoms
+ flaxseed x4 + tau x1

8.43pm: L arginine x2 + silica x1
+ ashwag x1 + Brewers yeast x6
Pneumonic + vapor rub wand neck
no boot worn today
Pneumonic little swell under lip
and nose

Harry Potter

9.50pm: yoga relaxation +
omega3 x4 + ashwag x1 +
Saw Palm x1

11:27pm: Omega3 x3 +
flaxseed x3 + Brewer yeast x6
+ Charcoal x1

1:34am (March 30 2019)
Omega3 x4 + Lira x1 +
Brewer yeast x6

2:40am: Omega3 x4 + Sage x3
+ menopause x1 + Brewer yeast x6
+ Charcoal x1 + shtp x1 + ashwag x1
+ evenfrm x1 gum sore +
neck stiffness

5:07 am: even prim oil x1 +
Omega 3 x 3 + sage leaf x 3 +
Brewer yeast x 6

6:02 am: Omega 3 x 3 neck
clavicle stiffness + pain

8:49 am: uva x1 + Brewer yeast x 6
+ omega 3 x 3 + even prim x1 + Silica x 1
amoxicillin x1

→ 1:35 pm
12:01 pm: even prim x1 +
L-arginine x 3 + Brewer yeast x 6
+ charcoal x1 + sage x 3 +
uva x1 + Omega 3 x 3 + silica x1

2:21 pm: Brewer yeast x6
3:29 pm: charcoal x1 + flaxseed x1

10:35am: St John's wort x1

11:45am: ever prim x2 + tau x1
+ Brewers yeast x6 + amino son x1
+ 5HTP x1

12:14pm: Brewers yeast x6

1:44pm: flaxseed x1 + Brewers yeast x6
+ Hydro collag x1

2:44pm: menopace x1 + Brewers yeast x6
+ Calcium D K x1 + ever prim x2

3:22pm: Omega 3 x1

4:14pm: Omega 3 x1

4:38pm: Saw Palmetto x1

Harry
Potter

6.48pm: Brewer yeast x6 + Sage x2 + Uva x1 + Omega3 x1 + even prim x1

7.30pm: Omega3 x 2

8.52pm: Calcium D K x1

10.34pm: Brewer yeast x6 + even prim x2

11.03pm: ashwag x2

12.57am [April 2019] Brewer yeast x4 + ashwag x1 + even prim x2

4.54 am: Sage x 3 + Brewer yeast x6 + omega3 x1 + ashwag x1

6.26am: 5HTP x1 + milk thistle x1

10.23am: Omega3 x3 + Ura x1

10.36am: Sage x3

2.29pm: Sage x3 + Eves prim x1 + Omega3 x2

4.01pm: menopace x1 + Sage x3 + Shtp x1 + Eves prim x2 + omega3 x1

6.30pm: flaxseed x2

10.16pm: Saw Palmetto x1 + Sage x3

10.40pm: down the stairs to pick up my Graze snacks + Brewes yeast x6

10.50pm: Shtp x2

Harry Potter

bus 266 → petrol station →
£60 cash machine → £25 elec £25 gas
→ bus 266 → prescription (DR) Deb
→ pepper, punch files, plastic cups
→ bus 266 (raining) → H&B
annovious, flaxseed 120, Turmeric tea,
green tea, Brewers yeast
→ Iceland Home Delivery
Blackcurrant capriSun, orange capriSun,
coke, binbag, eggs, potatoes, water
→ bus 266 → Home

Harry
Potter

(April 2 2019) 2am: evenprim x2
+ omega3 x1 + sage x 3

3am: Cranberry x6 + anal itch

3:15am: ashwag x1

10:07am: Urax1 + flaxseed x3
+ omega3 x1 + Brewers yeast x5

Home by 2pm

2:51pm: flaxseed x 3
+ Brewers yeast x6

Harry
Potter

Brewer's yeast is a key
herb to heal lots of illnesses
Six tablets every half hour
or hour x 6 - 10 times a day
with other Herbs will nullify
Cancer cells very quickly - and
within 3 weeks

Harry
Potter

Brewer's yeast is a key
herb to heal lots of illnesses
Six tablets every half hour
or hour x 6 - 10 times a day
with other Herbs will nullify
Cancer cells very quickly and
within 3 weeks

3.

Sp

5.

S.o

So
in t

6.1

ter

6.41p.

8.08

+ top

tanx

3.51am: Brewers yeast x6
(debittered)

5pm: ashwag x 1

5:25pm: amino1500 x 2 + omega3 x 1

5:44pm: Brewers yeast x6

So cold today but sunny
in the afternoon!

6:18pm: Brewer's yeast x6
+ everprim x 1 + ura x 1

6:41pm: Rainy evening

8:08pm: omega3 x 1 + l-arginine x 2
+ toptail wash + feet +
tan x 1 + ura x 1 + Brewers yeast x 6

10.05pm: Brewers yeast x 6 + charcoal x 1

10.29pm: flaxseed x 2 + amino15 00 x 1

10.34pm: omega3 x 1

11.16pm: Brewers yeast x 6 +
ashwag x 1

3am [April 3 2019] flaxseed x 1
+ pepper feel cough like +
uva x 1

3.17am: Brewes yeast x 6 + omega3 x 2
+ charcoal x 1 + ever primoil x 1

4.13am: Brewes yeast x 6 + a x 1
+ 5htp x 1 + charcoal x 1
+ amino1 00 x 1 + flaxseed x 1
+ ashwag x 1

Harry
Potter

10.05pm: Brewer's yeast x6 + charcoal x1

10.29pm: flaxseed x2 + aminoisoux1

10.34pm: omega3 x1

11.16pm: Brewers yeast x6 + ashwag x1

3am [April 3 2019] flaxseed x1
+ pepper feel cough like +
uva x1

3.17am: Brewer yeast x6 + omega3 x2
+ charcoal x1 + even primoil x1

4.13am: Brewer yeast x6 + tau x1
+ 5htp x1 + charcoal x1
+ aminoiso x1 + flaxseed x1
+ ashwag x1

5:35am: amino1000x1 +
Brewers yeast x6 some gum swell

5:49am: Bathroom + Omega 3 x2
+ salt water mouth rinse

11am: Brewes yeast x6 + iron x1
+ charcoal x1

12:05pm: Brewes yeast x6 + amino1000
x1

12:37pm: could lift left leg (bad leg)
up to the bedroom window seal but
not right (good) leg

12:43pm: Brewes yeast x6
+ silica x1

Harry Potter